WHEN THE FAE CAME

Martin Thomas

When The Fae Came

The moral rights of the author have been asserted by them.
© Martin Thomas 2026

10 9 8 7 6 5 4 3 2 1.

All rights reserved
No part of this publication may be reproduced, stored in a retrieval system, or transmitted in any form or by any other means (electronic, mechanical, photocopying, recording or otherwise) or used to train any artificial intelligence technologies without the prior written permission of the author. Subject to EU law, the author expressly reserves this work from the text and data mining exemption.

Previous Publications by the Author:

UFO – Friend or Foe?	August 2025	978-1-918045-21-5
We Own 29% - ET Has the Rest	September 2025	978-1-83709-256-7
Earth's Alien Syllabus	November 2025	978-1-83709-274-1
The UFO, E.T, Alien Trilogy	January 2026	978-1-83709-292-5

Website Address: MartinThomasAuthor.com

Individual copies of these books can still be acquired via my website.

When The Fae Came

Dedication
This work is dedicated to the BLAZE Television Channel for making me think, to Google for helping me find information and to Wikipedia for denying everything.

Malediction
I call down a blight on the AIs running on Google and Wikipedia. They make it harder to find anything which is not main-stream.

When The Fae Came

When The Fae Came

Table of Contents

Preface vii

1 WHAT IS GOING ON? 1
   a   I Get Convinced UAPs Are Real 1
   b   What Are The "Others" Doing? 2
   c   The Fae 3

2 THE FAE IN THE BRITISH ISLES 6
   a   Sightings Evidence 6
   b   Extracting The Relevant Data 8
   c   British Fae 10
   d   Where Did They Come From? 11
   e   The Alternative Story 13

3 WHAT FAE ARE OUT THERE? 15
   a   What Are My Sources? 13
   b   Species & Tribes 16
   c   Dwarfs 17
   d   Elves 21
   e   The Little People 27
   f   Goblins 32

4 OTHER FAE 36
   a   General 36
   b   The Fae and The Sea 37
   c   Little Green Men 38
   d   Animal-Like 40
   e   Vaguely Humanoid 41
   f   Difficulties in Identification 45
       I   General 45
       II   The Curupira 47

	III	A Creature with a Trunk	47
	IV	Small Grays	49
	V	The Hairy Fae	49
	VI	Little Green Men	52

5 THEY'RE EVERYWHERE 55
   a   General 55
   b   The Racing Certainties 56
   c   The Runners-up 58
   d   The "Also-Ran"s 58
   e   Limitations 58

6 WHY ARE THEY HERE? 60
   a   Elves & The Little People 60
   b   Goblins 61
   c   Dwarfs 62
   d   Other Fae 62

7 SUMMARY & CONCLUSIONS 63
   a   Summary 63
   b   Conclusions 65

APPENDIX 1   Sightings in the British Isles 67
APPENDIX 2   Illustration of Various "Others" 71
INDEX 97
REFERENCES 101

	Title	Page
Figure 1	Heights of Dwarfs with a Small Gray for Comparison - Dwarf, Gnome, Small Gray	18
Figure 2	Illustrations of various Elves –Leprechaun, Brownie, Elf & Pixie.	23
Figure 3	Examples of Little People-Sized Fae: Mexican Chaneque, English Fairy, Javanese Jenglot	31
Figure 4	Examples of Goblin Family - Goblin and Hobgoblin	35
Figure 5	Illustration of Types of Grays	51

PREFACE

Millions of people believe that there are mysterious objects flying in the sky. They were originally described as Unidentified Flying Objects – hence UFOs – but have more recently been termed UAPs – Unidentified Anomalous Phenomena. As most humans spend their time on the land, there are many more sightings of UAPs there, than have been associated with maritime UFOs, which have been termed Unidentified Submersible Objects (USOs). The abbreviation UAP is now used to cover both UFOs and USOs.

The term "Alien" has been used to describe any occupants or manufacturers of UAPs. However, this word has many pejorative associations – it means basically "not one of us" and is frequently used to describe immigrants, whose presence may be illegal, or who are disliked because of their race, colour or creed. I intend to use the word "Others", perhaps not an ideal term, but at least it avoids many of the pre-assumptions inherent in traditional terms.

In my previous work[1], I looked at the information available on the sightings of UAPs, and particularly of "Others", to try to understand what they were doing here. What stood out was that the various UAP researchers seemed to concentrate on sightings of "Others" that were as big as or larger than Humans. In extreme cases, they seemed confident to distinguish between a 2.0m tall Nordic called an Apunian, and a 2.0m tall Nordic called an Arian.

Annoyingly, sightings of smaller "Others" were dismissed as of little interest, or even cautioned against, on the basis that it was dangerous to interfere with spiritual beings. Large

"Others" were considered sexy, whilst small "Others" were a turn-off. Of course everyone considered they knew everything about those ambiguous creatures the Small Grays, but the word dwarf was used all too often to describe anything smaller than them, so ignoring the major contribution which Dwarfs have made to Human development.[2]

I discovered colonies of small "Others" called Chaneques or Aluxes in Mayan Mexico[3], vicious Jenglot in Indonesian Java[4], and small Elf-like "Others" in New Brunswick in Canada.[5] This was sufficient to convince me that small "Others" need to be looked at far more closely.

Chapter 1
WHAT IS GOING ON?

a) I GET CONVINCED UAPs ARE REAL

Ancient clay tablets[6] found in Sumeria dating back thousands of years suggest that Humans were visited by strangers called the Anunnaki, a race of giants that they worshipped as Gods. When viewed with a modern eye, these would probably be considered as some form of Extra-Terrestrial.

In my first book[7], I reviewed a wide range of UAP sightings to get a feel for their likely veracity. I soon excluded any sightings where there was only one witness, on the basis that I was not able to go into their possible motivations, whether honest, dishonest or delusional. Then, I eliminated mysterious cloud formations and single unmoving points of light, on the basis that they could so easily be miss-identifications.

What remained was a group of sightings which could not be explained away so easily. If it wasn't so worrying, it would be comical: there were desperate attempts by officialdom to account for these, including such fantasies as marsh gas burning[8] high in the air rather than getting diluted as it rose, or the highly improbable chance coincidence[9] of a meteor and an earthquake in a non-earthquake zone.

Then we have threats by mysterious agencies to witnesses and their families, either of violence or of destruction of their careers.

Even if the evidence of these remaining sightings were not so convincing, the desperate attempts at denial would probably have convinced me of the reality of UAPs and "Others".

b) WHAT ARE THE OTHERS DOING?

In my last book[10], I showed that there are a number of "Other" species who have established small colonies on Earth, together with many bases.

President Dwight Eisenhower made a grave mistake when he signed a deal[11] with an "Other" race known as the Small Grays[12]. In return for information about "Other" technology, he allowed them a base in the US and granted them permission to take cattle and humans for experimental purposes, provided that the humans were returned unharmed, and they limited the numbers that they took.

The Small Grays, once they were comfortably installed, have broken every part of the agreement, and the US can do nothing about it. They are now considered responsible for most of the human abductions around the world.

c) THE FAE

A sizeable proportion of the world believe that, in addition to the physical world in which we live, there is a magical realm which overlaps with ours, so that we can sometimes see the denizens of this realm[13]. They are mostly smaller than us. Some call the whole family of supposedly magical beings Elves[14] with Fairies as a type of Elf, and some call them Fairies[15] with Elves as a type of Fairy. Within the British Isles, they have a variety of names, depending on the cultural traditions of the area. Often there is more than one name:

Ireland	The Good People	People of the Mounds
Isle of Man	The Little People	Themselves
Scotland	The Peaceful People	The Fair Folk
Wales	The Fair Folk	The Blessing of Mothers
England	The Fae	

These names seem designed to flatter them, to keep on their good side. I am going to use the English name; The FAE.

They are claimed to vary from Fairies of less than 10cms tall through Gnomes, and Brownies to Dwarfs and Elves nearly 1.5m tall.

Do you believe that The Fae are magical? As an engineer, I believe in an ancient principle called Occam's Razor. The fundamental tenet of this is that you should always consider the easiest explanation when looking at a problem, and only escalate the complexity if that doesn't work.

If there are some species on Earth which have very strong magical capabilities, then there should be other species with weaker abilities, and virtually every creature on Earth should have some. This does not seem to be the case.

What capabilities do Large "Others" possess? Some can fly, some can vanish in plain sight, some are telepaths, and some can impose their will on we poor humans. We accept that they can achieve this with their technology

What capabilities do these small, supposedly-magical "Others" possess? Some can fly, some can vanish in plain sight, some are telepaths, and some can impose their will on we poor humans

What is the difference?

I believe in the late Arthur C. Clarke who famously said:

> Any sufficiently advanced technology is indistinguishable from Magic.

I am not like the White Queen in Alice through the Looking Glass who could "believe in six impossible things before breakfast". I can only cope with one at a time.

I remain certain that UAPs and "Others" exist, even if they have been in our folklore as far back as human memory can stretch. I hope that I managed to convince some of my readers in my first book.

In this book, I propose to look at the world of the Fae from the point of view of someone who doesn't believe in magic, but does believe in advanced technology.

I will take the Fae population in the British Isles as an example of a closed sample area, where incomers need to make a determined effort to enter it, and where sightings have been recorded for many years. I will then look further afield to see how representative this sample is.

I accept that there are sincere skeptics who cannot believe in UAPs and "Others", but I invite them to read further to see whether some strange events are better explained in my interpretation, than in the "Official" version or the "Romantic" version. Otherwise, just read and enjoy it as a fantasy.

Chapter 2
THE FAE
IN THE BRITISH ISLES

a) SIGHTINGS EVIDENCE

In his book "UFOs, Humanoids and Strange Phenomena of England[16], George Mitrovic fills 575 pages with descriptions of events encountered between 1614 and 1999. He describes more in the 232 pages of his book "UFOs, Humanoids and Strange Phenomena of Ireland, Scotland and Wales[17], dating from 1765, 1613 and 1656 to 1999 respectively.

These two books describe:

- Loud explosions followed by earthquakes. Could these be "Other" species excavating underground and underwater bases?
- Strange things falling from the sky such as coloured rain, lumps of ice, lumps of different types of rock, and falls of fish, frogs and other creatures.
- Objects flying in the sky, including descriptions and details of their activities.

- Various anomalous creatures such as multiple sightings of Plesiosaurs, Puma, Yeti, Robots and Men in Black.
- Cryptids such as Dogs with Wings, Giant Owls, Large Dogs with Pig's Snouts. Creatures looking like Werewolves, Giant Winged Creatures looking like Dracula, and Large Talking Foxes.
- Large "Others" with heights ranging from 1.5m to 9m, skin colours of white, green, blue and black, various amounts of body hair, with and without heads or faces, and even scaled.
- Small "Others" with heights ranging from less than 10cms up to 1.5m, all skin colours and textures, and different amounts of hair.
- The ghostly appearance of humans apparently from history, such as Cavaliers, Air Raid Wardens, Boys with Gasmasks, Ladies in Edwardian Dress, Ladies in Bonnets, and RAF Personnel.

In Scotland, there are so many recorded sightings of Nessie-type Sea Monsters in their lochs that, coupled with records of sightings in Norway, Sweden, South Africa, South America and Australasia, I can only suggest that there is a breeding population of Plesiosaurs in the Earth's southern oceans, possibly with the lochs, lakes and fjords of the north acting as nurseries for their offspring.

You may wonder why I have chosen to split the "Other" sightings into two groups. It is simply that, whilst we can accept that any "Others" from off-Earth must have superior technology to us, we seem content to accept that creatures shorter than 1.5m tall must be magical.

There are many, possible mythical, species which are reputed to have magical capabilities, most of whom are known as "The Fae". In this chapter, I propose to look at their reported sightings to see who are present.

b) <u>EXTRACTING THE RELEVANT DATA</u>

In the Table in Appendix 1, I have selected sightings of small "Others" on a region by region basis, trying to spot if there are any concentrations.

It is interesting that many of these sightings are by young people so, are the Fae attracted to them, or are young people more capable of spotting them? Of course, young people can be more imaginative too. This means that we have to be careful in accepting their sightings at full face value. However, in all modern cases their reports are followed up by an investigator, and attempts are made to eliminate the fanciful.

One set of sightings is very confusing: There are a good number of reports of "Others" looking like monks wearing cowls, and often walking in line, but there do not seem to be any estimates of their heights. Are they Fae or not?

When someone is called small, what does that really mean? It could mean absolutely tiny, like an adult's thumb, or it could mean a bit smaller than the speaker. This gives a range of about 0.1m (4") up to say 1.6m (5'3"). It is clear from the Table that witnesses are rarely willing to be too specific about heights. The figures look accurate when the metric measurements are considered, but these witnesses are British! Two-thirds of the reports are for 1, 2, 3, 4 or 5 feet tall, with only one third willing to be more specific.

This will make it difficult to use heights to differentiate between species of Small People, at least in the British Isles.

Outside these criteria, there is an observable increase in sightings of tall "Others", in all their variety.

There are also many more claims of abduction, where a landed UAP is sighted and the witness knows nothing more, but finds scars and marks on their body later. There is a corresponding huge jump in sightings of 1.5m "Others", which is a bit above the height of typical Small Grays, the species most blamed for abductions[18], but does likely reflect their presence

One feature which stands out in the data is that, if you compare the number of sightings of Fae in a Region before 1945, excluding 1.5m "Others", with those after 1975, some increase and some decrease.

There seems to be a trend for these sightings to increase inland and towards the north.

Is this a real change, or are those in the west and east simply so blasé that they no longer bother to report such sightings?

Either way, there is still a lot to be learnt from this Table. If we look at the pre-1945 sighting reports, despite the bias towards 1, 2, or 3 feet, we can condense this data and see several things:

- In the Isle of Man, there is a cluster of sightings at 3 foot,
- In Scotland, there is a cluster at 2 foot,

- In the North West, there is a cluster at 2 foot,
- In the South East, there is a cluster at 3 foot,
- In South Wales, there is a cluster at 1 foot 8 inches or perhaps 2 foot,
- In North Wales, there is a cluster at 3 foot,
- In the East of England, there are clusters at 1 and 2 foot,
- In the East Midlands there are clusters at 1 and 2 foot,
- In the South West there are clusters at 1, 2 and 3 foot,
- In Ireland, there are clusters at 1, 2 and 4 foot.

Before we can see the possible implications of this, we need to identify what are the heights of the possible Fae in the British Isles.

c) <u>BRITISH "FAE"</u>

There are a number of Fae names which come up in British folklore again and again[19, 20, 21, 22], and it is likely that they do refer to the local species.

These are:
- Brownie – 75cms (2'6")
- Dwarf – White – 90cms (3'0") and Brown (Duergar) – 45cms (1'6")
- Elf – Scotland 1m65 (5'6") and England 1m20 (4'0")
- Fairy – 15cms (0'6")
- Little People – 30cms (1'0)
- Gnome, Knocker – 30cms (1'0")
- Goblin, Red Cap – 60cms (2'0")
- Hobgoblin – 45cms (1'6")
- Leprechaun – 75cms (2'6")
- Pixie – 60cms (2'0")

It is possible that Trows (small Trolls) may also be present.

The heights of the various species are best guesses, based on the various lists of sightings.

In Ireland, there are two apparent clusters. In the cluster at 1 Foot, the sightings reports include 8 which identify Fairies. The smaller cluster at 4 Feet includes 3 White Dwarfs. Otherwise, the sightings include 4 Leprechauns of varying height, one hairy creature that might be a Brown Dwarf (Duergar) and one Hobgoblin in his rags.

In Scotland, apart from all the sightings of Lake Monsters, in many different lochs, there is a cluster of really lively Hobgoblin-like Fae living on the moors. There would also appear to have been something happening at Comrie in Perth & Kinross, where there are repeated earthquakes and large flashes of light. It is known as the "earthquake capitol of Scotland". Could the Fae be building something there? Sadly, we also have the first two sightings of Small Grays (1.3m) in the British Isles, as they slowly move in.

In the Isle of Man there was one sighting of a tiny Fairy and repeated sightings of 3 foot high Fae, who always seem to be dressed in red. Could these be small Elves?

Wales has a cluster of Dwarfs in the north. White Dwarfs tend towards mining areas, such as North Wales and the West of England, where there are tin, copper and gold mines. It may be that they are also present in the coal mines in Yorkshire, but none have been reported. Perhaps they stay underground where there are coal mines, as their skills would never be needed above ground. In the South Wales coal mining area, there has been only one sighting of a White Dwarf, and a few Hobgoblins.

In the West Country, there is a cluster of tiny Fae with 5 Fairies, and one hairy Little Person, and a larger cluster including sightings of 7 Pixies. There are also sightings of 2 White Dwarfs and one Goblin.

In the North East, where there is claimed to be a Dwarf hideaway in Cragside, there are sightings of short Fae, who could be Brown Dwarfs, also called Duergars.

d) <u>WHERE DID THEY COME FROM?</u>

Some people claim that the Fae live in a parallel dimension, which overlaps with ours. The White Queen must be splitting her sides with laughter – not one impossible thing (UAPs), not two impossible things (Magic), but three impossible things before breakfast! I am going to stick with just one impossible thing.

In Ireland there is a Manuscript called "The Annals of the Four Masters", which describes[23] how the Fae, the Tuatha Dé Danann, arrived on the Connaught coastline a bit earlier than 1900 BCE, in a great mist, and then burned their ships. UAPs, landing in a mist or fog, have been experienced in many places around the world[24].

They displaced the then rulers of Ireland, the Fir Bolg, and themselves ruled Ireland from 1879 BCE to 1700 BCE before they were conquered by invading Celts. These allowed the Fae to live, provided they dwelt underground.

This is the only recorded landing in the British Isles, but the Dwarfs apparently appeared in Anatolia about 3000 BCE,

When The Fae Came

and led early humanity through the Bronze Age and the Iron Age[25]. They could have arrived in Britain at the time of the British Bronze Age in 2000 BCE.

In February 1995, a lorry-driver in Mexico, claimed to have found a human-like "Other", 0.30m tall, sitting in his passenger seat[26]. This individual said that he lived in a colony at the bottom of Lake Chapala, which is near to Guadalajara. It is a very big lake (75km x 25 km), but shallow, with a maximum depth of 9.0m. He claimed that he was descended from the occupants of an UAP that crashed thousands of years ago. This would certainly fit in with the presence of Alux, the Little People described in Mexican folklore, that they called Chaneques.

Apparently, there is no record of Kobolds in Germany prior to the 13th century[27], so perhaps that is when they landed.

There is a modern instance too. A race of extremely tiny (7.5cms) "Others" was seen to land in Malaysia[28] in about 1960, and they are thought to be still there.

e)   THE ALTERNATIVE STORY

So, we can come up with an alternative history of the Fae, which requires only one impossible thing to be believed, before or after breakfast - UAPs.

The Fae comprise many species of "Others", which have arrived on earth, probably to set up colonies, at different times over the last few millennia. Their landing places are widely scattered, and they have probably spread to cover those parts of Earth which they didn't reach to begin with. They brought with them their technology which permitted them to travel from their own planet, together with devices to let them fly, disappear, cast illusions perhaps using holography, and to read and project thoughts.

The next question is whether all the Fae, sighted around the world, belong to the same few species seen in the British Isles, or whether there are many more to consider.

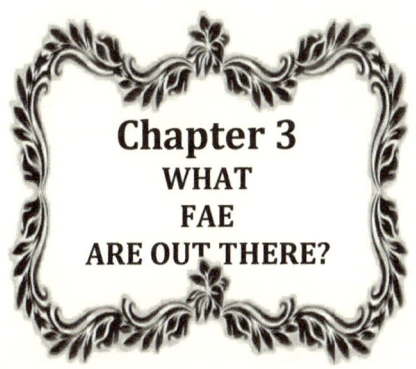

Chapter 3
WHAT FAE ARE OUT THERE?

a) <u>WHAT ARE MY SOURCES?</u>

There are a number of works on Fairies and other Fae. The difficulty is that they sometimes use different names for the same species, some perhaps understandably based on the local name for them, but some rather fanciful.

There is a further complication in that an individual creature may be described very differently in different works, and given totally conflicting attributes. For example, the Duergar or Brown Dwarf is described both as "sometimes malevolent" and as "highly respected".

Rather than using the list of references each time to identify a particular source, I will list the authors here, and cite them only once in the back.

    i) The Encyclopaedia of Fairies- Theresa Bane[29]

    ii) Here Before Us - Giants & Little People - Peter Netzel[30]

    iii) Briggs's Dictionary of Fairies – Briggs & Greening[31]

iv) The Book of Fairies – Frances Melville[32]

v) Field Guide to the Fae – Nancy Arrowsmith[33]

vi) The Little Encyclopaedia of Fairies – Ope Opanike[34]

vii) Magic Myth and Mystery of Dwarfs – Virginia Loh-Hagan[35]

viii) The Fairy Census 1 – Britain & Ireland. S R Young[36]

ix) UFOs, Humanoids and Strange Phenomena of England by George Mitrovic[37]

x) UFOs, Humanoids and Strange Phenomena of Austria, Belgium, Estonia, Germany, Holland, Kalingrad, Latvia, Lithuania, Luxembourg, Poland & Switzerland by George Mitrovic[38]

xi) UFOs, Humanoids and Strange Phenomena of Andorra,
   Gibraltar, Spain & Portugal by George Mitrovic[39]

xii) UFOs, Humanoids and Strange Phenomena of Africa, Asia and the Middle East, P105. by George Mitrovic[40]

xiii) Boggarts, Brownies, Hobs and their Goblin Kin by Stephen Rae[41]

xiv) Fairies by Janet Bord[42]

Whenever I need to cite a reference in the rest of this chapter, I will simply use the Roman numeral given above.

b) SPECIES AND TRIBES

When we look at the Human Species, we see many differences in the races around the world. We have different skin colours, hair colours, heights and eyes. Some differences, such as our height and build, can evolve quite quickly - say in a thousand years or less. Other differences, such as skin colour would take far longer.

The same applies to the Fae. Dwarfs can change in height and build but not in colour. Goblins' ears could not easily evolve from Dwarfs' ears, nor from Elves' ears. I think it unlikely that the tiny winged Fairies could evolve from Elves.

We must be careful to bear in mind that different species can undertake the same roles without having to be branded as the same species. Either a Brownie or a Hobgoblin could be welcome in a home.

c) DWARFS

I) General

So let's start with Dwarfs. There is general agreement that they are short, sturdy and bearded. The consensus seems to stop there. Beyond that, it is a matter of picking what seems right. There are at least 3 different sorts present in the British Isles. One book (vii) tells us that there are four types –

- white (1.0m),   Peaceful
- brown (0.6m),   Can cause trouble
- black and   Evil
- red   Starts fights

– although there are no reported sightings of red or black Dwarfs in the British Isles. The 0.6m Dwarf has also been described as a black Dwarf (iii), but those limited sightings in Britain suggest that its behaviour is not as bad as that.

The Gnome (0.3m) is simply a smaller version of a Dwarf (i).

These three could all have come to Britain from the Mediterranean area, or they could even have evolved from common stock since arriving.

Figure 1 shows illustrations of Dwarfs and Gnomes with a Small Grey for comparison.

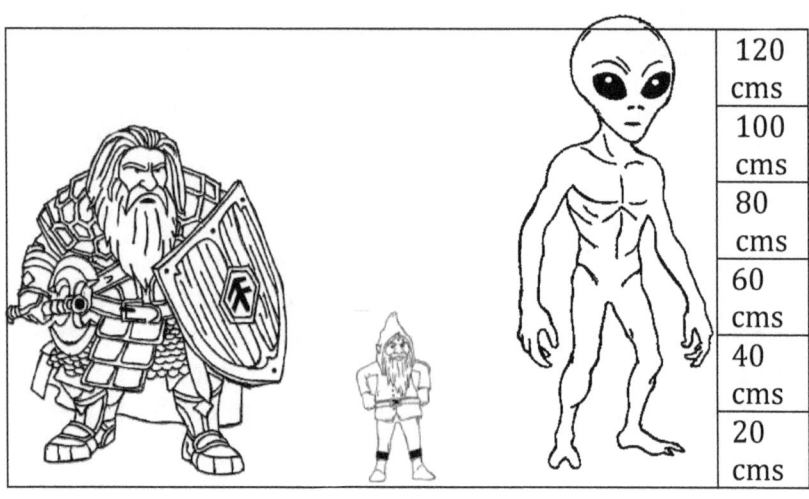

Figure 1 Heights of Dwarfs with a Small Gray for Comparison
Dwarf, Gnome, Small Gray

II)   Pre-History

There is no definitive proof that Dwarfs came from Anatolia, or that they began the Bronze Age. However, it appears to have started there about 3,000 BCE[43]. Indirect evidence comes from ancient Anatolian houses, particularly those at Çatalhöyük, which were too low for humans, closely packed, and had flat roofs that served as streets[44]. In Eastern Iran, there is another village sized for Dwarfs, called Makhunik[45].

III)   Dwarfs World-Wide

In searching my list of references given above, I have been disappointed to find that the phrase "Dwarf-like" seems to be interchangeable with the word "Small", so that most of them are useless, and very few of these relate to bearded folk.

Belgium before	There are reports of Dwarfs in spacecraft and after the Second World War.(viii)
Egypt	There is an ancient relief-carving of a deity called Bes, who definitely looks like a Dwarf, right down to the beard (ii).
France	In the story of King Loc, his description makes him sound like a Dwarf, and his followers like Gnomes (ii).
French and	Swiss Mountains, there is a hairy Gnome (vi) called a Burbegazi. This could easily be a necessary adaption to the cold.
Germany	One book says that one of the three types of Kobold is an ugly mine-dwelling Dwarf, so could be a Brown Dwarf (ii). In another book (iv), a tribe of white Dwarfs is mentioned which are noted metalworkers. In Cologne they talk about a House Gnome (vi) called a Heinzelmännchen. However, as one would

	hope, there are tales of Dwarfs and Gnomes going back as far as 1644 (viii), and running right through to 1998, with rumours of a Dwarf colony somewhere underground perhaps in the Black Forest.
Greece	They seem to have three Dwarf villages (ii) - Phrygian Dactyls near Mt Ida, Kaburol Dactyls which is associated with the sea, and Rhodian Dactyls where there are noted blacksmiths. They also have a Gnome called a Krallikantzaros, the size of a small child.
Indonesia	They have an Orang Pendek (ii) which is 1.3m tall with broad shoulders, long arms and red hair. No mention of a beard, but it sounds like a Dwarf.
United States	In South Carolina there is a Yehasuri (vi) which is probably a brown Dwarf, judging from its reputation.
Native Americans	have a long history of sightings of Fae (ii), but few of the creatures sound like Dwarfs. In Detroit there have been sightings of a Goblin with red fur called the Nairn Rouge which is apparently more common in Ottawa (ii).
Norway	They have a grumpy Dwarf, better known as a Duragar.(ii), and a Gnome called a Nissi (vi). Their mythology contains a large number of Dwarfs.
Switzerland	There have been sightings of both Dwarfs and Gnomes.(viii)

Poland   There have been reported sightings of Dwarfs and Gnomes since 1894 (viii)

In my book "We Own 29%, ET Has the Rest[46]", I looked at records of coastal sightings of "Others" and, amongst these, there were 92 sightings of Dwarfs. These occurred mainly since 1945. The majority of these were either in Brazil and Argentina, in Northern Italy, or in Australia.

I was able to demonstrate that there was a Dwarf colony somewhere near Genoa. On the basis of these other sightings, I would suggest that there is a Dwarf base in Australia, and probably a couple in South America.

In addition to the probable Dwarf landing in Turkey millennia ago, there was almost certainly another more recent one in central South America, probably now with a colony being developed.

The Dwarfs today seem to have withdrawn from Human contact in most places, probably retiring to their colonies. They are really only spotted now in South America.

As I have said before, UAP and "Other" hunters seem to have minimal interest in the Fae, particularly Dwarfs.

d)   ELVES

I)   General

The definition of Elves seems to be fairly loose. I am going to describe them as slender and graceful with mainly human features but with pointed ears. Those in Scotland are apparently as tall as Humans, but the tallest seen in the rest of the British Isles are seldom higher than 1.50m.

When The Fae Came

Within that definition we also have Leprechauns (1.3m), Pixies (0.5m) and Brownies (1.0m). The Elves landed their spacecraft in Ireland in 1900 BCE, and burned their ships[47]. They ruled there until 1700 BCE, when they were forced, by the invading Celts, to take to living underground.

Since then, they have spread out across the British Isles, and probably much further afield.

When The Fae Came

Figure 2 Illustrations of various Elves –Leprechaun, Brownie, Elf & Pixie,

II) Elves World-Wide

Elves have spread world-wide, and appear in many forms:

Austria (v)	Wichln
	House Elf
	Weiss Frauen
	Aguane 1-1.5m
Balkans(vi)	Ursitary one of 3 fates
Brazil(ii)	Caipora
	Ikals 1.0m hairy
China(ii)	Drope & Ham people 1m30
Czech(v)	Hey-Hey men – shape shifters deserted areas
Denmark(v)	Ellefolk
Finland(vi)	Haltya
France(v)	Dames Vertes tall and beautiful
	Korred 0.3 -1.0m Brittany, cats claws
	Fountain women 0.6m
	Korrigan & Namignak
Germany(ii)	Zwerg
	Kobold 0.3 - 0.6 m household
	Weisse Frauen now rare
	Heinzelmännchen 0.3 – 1.0m
	Red Hat House Elf
	Görzoni
	Bushfrauen

When The Fae Came

Ghana (vi)	Asamanukpai
Greece (v)	Callicantzarol 0.6 – 1.0m
	Mocrae 1.0 - 1.6m Tells newborn destiny
	Nereides young slender
Hawaii (i)	Menehues - a Gnome.or Elf widespread in Polynesia
Holland(v)	Alven - almost invisible
Iceland(v)	Fylgiar
Italy(v)	Fate 1.6m wood & water elves
	Giana 1.6m Sardinia
	Folletti 0.3 – 1.0m
	Massariol 0.3m small farmer
	Caccavecchi (vi)
Japan(vi)	Kigemana
	Thussels (vi)
Lithuania(vi)	Kaukai
Mexico(ii)	Choneque 0.3 -0.6m
New Zealand	Patupaiarehe (v)
Norway(v	Ellefolk
	Fossgrim (vi)
Poland(viii)	Reported sightings
Russia(v)	Leshie & Kisonki forest Height varies
	Domoviye Hairy House Elf 0.6 – 1.0m
Senegal(vi)	Bachna Rachna,
	Yumbo
Slavs(v)	Domovie Hairy House Elf 0.6 – 1.0m

Spain (v)	Duendes House Elf 0.6m
	Xana
Sweden (v)	Elfor
	Nissen in houses
	Fossgrim (vi)
Switzerland (v)	Erdluitle 0.5 – 1.0m Control weather
	Servan 0.3 – 0.6m home sprite
US Native Americans(ii)	Crow - Nirumbee 0.6m
	Sioux – 0.4m
	Iroquois – Gahonga, Gandayah, Jegah, Ohdawa (vi)
	Wisconsin – Winnebago water sprite
	Cherokee – Yundi Tsurindi 0.6m
	Choctaw – Forest Dweller 0.6 – 0.9m
	Comanche – Nunupi 0.3 – 0.45m
	NW Indians – Stick Indian in Forest 1.0 – 1.3m
	Menominee - Marneguesi ugly water sprite
	Cree - Marneguesi ugly water sprite
	Alaska – Inukin 0.15 – 0.45m hair covered
The Sea(v)	Klabautermannikin 0.3 – 0.6m
Everywhere	Fée - Vary widely

As you can see, Elves are just about ubiquitous. I'm sure that there are more around the Pacific Ocean which have not been recorded.

It is interesting how widespread Elves are in North America, where they are probably known to every Native American tribe, not just those listed.

It seems clear that, to be this widespread, Elves have either been on Earth a very long time, or have made multiple landings in addition to the one in Ireland. They seem to be happy to live in the countryside in groups, and sometimes as solitaries as in the case of house Elves. They are reputed to live in caves in the winter, and in temporary sites such as under logs in the summer.

I haven't come across any evidence of their building colonies but perhaps, if they have been here that long, they have finished such things long ago.

e) <u>THE LITTLE PEOPLE</u>

I) <u>General</u>

There were only 18 reported sightings in my main source for the British Isles (vii) where the Fae was shorter than 30cms, but as I have already suggested, this might be due to a lack of interest in the smaller "Other" species. In those cases where there is a more positive attitude (vii) there are over 500 reports.

As I've already said, many people consider Fairy and Elf as generic terms covering the whole range of Fae. To me, the phrase The Little People covers Fae under 30cms tall. The word Fairy is confined to winged Little People

If it is suggested that some glow in the dark, we will have to find an explanation which does not require the use of magic.

The one explanation, which could possibly fit with my wish to only have one impossible thing to swallow, is that it is based on the same principles that enable UAPs to fly in our atmosphere. A power source is needed, and it has been suggested that power is transmitted through the atmosphere from pyramidal generators such as the one in Mount Denali in Alaska[48,49] to power these craft. Fairies would only need the capability of drawing on this energy – not an impossible capability. Bio-luminescence is a well-understood phenomenon.

Where Little People have wings, they are probably small enough and light enough to be capable of flight.

II)   Little People World Wide

There are nowhere near as many sightings around the world, as there are of Elves. Illustrations of 3 types are given in Figure 3 below:

Australia(ii)	Nuku-Mai-Tore, who sit on branches like birds
Baltic (ii)	Barstukai – size of finger
Egypt(ii)	Green Lady, who hides in trees & protects paths

When The Fae Came

France(ii)	Farfadet, a furry sprite Fée, who make Fairy rings and dance in them Couri-Brittany Follet troublesome House Fairy(xiii) Korrigan – wings like a wasp
India(xiv)	Simla – 18cms green & blue
Malaysia(ii)	Landed in 1959
Mexico(ii)	Lux, locally called Chaneques, who are said to have crash landed thousands of years ago.
New Zealand	Patupairehe – fair skinned(ii) Pakepakola – fair skinned(ii)
Poland(viii)	3golden haired sprites in a tiny spacecraft
Polynesia	Kakamosu (vi)
South Africa	Abatura – so small they ride on ants.(ii)
Spain	Anjana (ii) There have been a few recorded Sightings of really tiny Fae only 4cms tall (xi)
Surinam	Bakru (vi)
W Africa	Aziza (vi)

In Roman and particularly Greek mythology, there are many minor gods and goddesses, most of them being of the average Human height of the time. This included the Nymphs and Naiads, and there were hundreds of them.

However, there were a few who are painted as children with wings – Cupid and Harpocrates were minor Roman gods. They were probably too tall to be classed as Fairies, but they could have been related to Pixies.

Nike, Iris and Psyche, minor Greek goddesses, were also portrayed with wings. Certainly Psyche was small enough to have butterfly wings, and Nike was shown sitting in Zeus's hand, so could have been related to Fairies.

These winged gods and goddesses were very few amongst the vast pantheon of their gods.

Amongst the hundreds of other nymphs, dryads, naiads etc, which were generally Human sized, there may have been a few Little People, but they are not identified a such.

I can't offer an explanation for the shortage of Little People sightings worldwide. Perhaps they are not being recognized for what they are, or they are not being seen, or they are simply not there to see.

It is possible that the Fairies and some of the other Little Peoples do not have the same level of intelligence as the other Fae. It seems likely that the Fairies, at least, are ruled more by their emotions. These could simply be pets of the Elves

Figure 3 Examples of Little People Sized Fae:
Mexican Chaneque or Alux, Fairy, Javanese Jenglot

f) GOBLINS

I) General

Goblins are best recognized by their prominent ears which stick out sideways. They are typically 0.6m tall. They have a reputation for being particularly nasty. Their one relation in the British Isles is the Hobgoblin which, in contrast, is generally welcome as a house dweller, acting in much the same way as a Brownie. They are shorter than a Goblin at 0.4m tall. They have the reputation that, unlike the Brownie who just leaves if offended, they will become a Boggart, and start to disrupt the household.

(A Boggart (vi) is a spiteful creature, which can make milk go sour, or things disappear and, at its most malicious, can kidnap children. An apology or act of kindness is needed to rectify matters)

In the British Isles, there are few recorded sightings of Goblins, mainly because they cannot be identified from the details given by the witnesses.

II) Goblins Worldwide

There are a number of different types of Goblin, principally identifiable by their typical ears. Those sighted include:

Andes(xiii)	Lives in caves. Blond with a beard
Arctic(xiii)	Mahaha – touch so cold it kills
Australia(xiii)	Yara-ma-yha-who – a bloodsucker Quinkins – Cape York – Fat bellied, big ears, pot bellied, kangaroo tails. Takes children

When The Fae Came

Chile(xiii)	Fiura – Very ugly female. Breath can kill Tranco – thick hair. Seduces women lost in woods Duende – house Goblin
China(xiii)	Yaogoy – malevolent
Cuba (xiii)	Cuije – a water goblin
East Africa	Adroanzi stalks at night (xiii)
France(v)	Lutin – a troublemaker. 1.3cms- 45cms . US too.
Germany	Biersal.1m Big round eyes, pointy ears A boozer(xiii)
Greece (xiii)	Kallikantzaroi – black, hairy, tail
Italy	60cms - Linchetti gives bad dreams (v) Mazzamurello – Scares the bad, lives in trees(xiii)
Japan (v)	Kappa 60cms – basically dangerous – a vampire Ashinaga & Tonage means good weather (v) Bakemono can cause illness (xiii)
Korea(v)	Dokkaebi – discarded household items

When The Fae Came

Mayan(xiii)	Alux – Either benevolent or malicious Often looks like an owl
North America(xviii)	Pukwudgie 90cms – Native Americans describe him as malevolent with large nose and ears.
Paraguay(xiii)	Karai Pyhare short & ugly, punishes animal killers
Phillipines(xiii)	Nuno – Lives underground. Peaceful unless disturbed Tiyanak – a trickster
Poland(viii)	1.0m Black with large pointy ears 0.4m Oval head, pointy ears 1.2m Large ears. Hairy body 0.8m Hairy body, wrinkly, pointy ears.
S Africa	Tokoloshe – Malicious (v)
Scandinavia	Mara – causes nightmares (xiii)
Slavic(xiii)	Likho - Skinny old woman, one eye. Bad luck Skritek – House Goblin Leshy – Wood goblin – punishes animal killers

Spain(xiii)   80cm with a thin head like a pointy light bulb,
Slanty eyes, wrinkly skin and pointed ears
Trasgu – horns, tail. Can enter house and break Things(xiii)
Duende - 60cms hairy house Goblin (xiii)

Figure 4 - Examples of the Goblin Family
Goblin, Hobgoblin and Lutin(US)

Chapter 4
OTHER FAE

a) <u>GENERAL</u>

Once again, in this Chapter, I have to reference a few books so many times, so I am going to list them here, and then only have to user the References section once for each. The books are:

i) UFOs, Humanoids and Strange phenomena of Bolivia, Brazil, Columbia, Ecuador, Guyana, Suriname and Venezuela by George Mitrovic[50]

ii) UFOs, Humanoids and Strange Phenomena of Argentina, Chile, Paraguay, Peru and Uruguay, by George Mitrovic [51]

iii) UFOs, Humanoids and Strange Phenomena of Andorra, Gibraltar, Spain & Portugal by George Mitrovic.[52]

iv) Amazing Encounters with UFOs in Central North America V1& V2 by George Mitrovic[53]
v) UFOs, Humanoids and Strange Phenomena of Austria, Belgium, Estonia, Germany, Holland, Kalingrad, Latvia, Lithuania, Luxembourg, Poland & Switzerland by George Mitrovic. [54]
vi) UFOs, Humanoids and Strange Phenomena of Africa, Asia and the Middle East by George Mitrovic[55]
vii) UFOs, Humanoids and Strange Phenomena of France by George Mitrovic[56]
viii) Amazing Encounters with Monsters and Other Mysteries of Australia, New Zealand, the Pacific and Antarctica by George Mitrovic [57]
ix) UFOs, Humanoids and Strange phenomena of Central America, the Caribbean and Mexico as well as the Atlantic Ocean. By George Mitrovic.[58]

In the rest of this Chapter, I will use roman numerals to refer to these books.

b) THE FAE AND THE SEA

In all the thousands of sightings recorded, there is one serious deficiency – there are pitifully few water-based Fae sightings.

The most obvious lack is records of Merpeople, whether Mermaids or Mermen with only one sighting in the British Isles and, world-wide, none mentioned in any of the records I have searched. It has been suggested that Merpeople belong to an Earth civilisation far older than we humans[59] but that elements of our race are attempting to wipe them out. I can find nothing to support this assertion, apart from the limited number of recent sightings.

These records do mention other water entities of various forms, such as Elves like the French Fountain Women, Italian Fate, Scandinavian Fossgrim and Native American Marneguesi, and Goblin kin like the Cuban Cuije.

The Mediterranean Sea does seem to abound with Nymphs and Oceanids. However, these "Others" are too big to be called Fae.

Overall, given how widespread the Fae are on land, it is difficult to believe that they have not extensively populated lakes, rivers and the sea.

c) <u>LITTLE GREEN MEN</u>

DESCRIPTION	COUNTRY	PAGE
(Little Green Man 1) Reported at 50cms tall, this little person is described as having green skin, large ears and large slanted eyes	Brazil (i) Brazil (i) Brazil (i)	94 209 211
(Little Green Man 2) Standing at 75cms, it has green skin and very long arms and protruding stomach and buttocks. Heads like pears.	Spain(iii) Estonia (v) France(vii)	53 50 144

(Little Green Man 3) A 1.0m tall entity with green skin and eyes/ears on stalks.	Spain (iii) France(vii) US Michigan(iv) Puerto Rico(ix)	220 38 1976 241
(Little Green Man 4) 1.0m. Dark Green Skin, with large eyes, pug nose, and arms at side. Rolls of skin on top of head.	US Ohio(iv) France(vii)	1974 177
(Little Green Man 5) 1.1m Pale green, long arms, 2 fingers, bow legs, eyes close together	US Iowa(iv) Indiana(iv)	1969 1955
Green Dwarfs About half of the reported sightings in Argentine describe them as green, but nowhere else in the world. Could it be that they are being miss-identified?	Argentina (ii)	

d) ANIMAL-LIKE

DESCRIPTION	COUNTRY	PAGE
(Octopus) Standing between 1.0 and 1.2m, this creature has 4 or more arms according to reports	Spain(iii) US Nebraska (iv) Puerto Rico(ix)	49 1984 202
(Frog1) 1.05m, thin wide mouth, right arm longer than left, claw-like hands, moves oddly. Orange eyes, small ears. Scaly skin	US Ohio (iv) US Wyoming(iv) Australia(viii)	1955 1971 99
(Frog2) 45cms. Smooth skin, striped green & brown. Frog-like, small hands, 3-toed feet. Webbed	US Illinois(iv) US Indiana(iv) Puerto Rico(ix)	1950 1955 220
(Dog Man) At about 1.0m, this creature has hair and a tail, appearing dog-like	Poland(v)	205
(Bird) Like a large black owl, this is big at 1.2m	Belgium(v) Puerto Rico(ix)	43 211

When The Fae Came

Chupacabra 1.2m Vicious teeth and claws, blood sucker. Ran into UAP	Puerto Rico(ix)	301
(Birdman) 20cms tall. Human legs, V-shaped head.	US Montana(iv)	1984
(Insectman) 1.2m. Green Compound eyes, big hands, rounded clubbed feet, 3 fingers	US Utah(iv) US Idaho(iv) France(vii)	1969 1997 220
(Kangaroo) 1.0m Like cross between Kangaroo and Cat. Pointy ears, light green. 2 short arms, big legs, antenna	Puerto Rico (ix)	247

e) <u>VAGUELY HUMANOID</u>

DESCRIPTION	**COUNTRY**	**PAGE**
Curupira This is described as being 0.6m tall with a face that appears to have a bad case of acne, and ping-pong ball type eyes, red hair and claimed backward-pointing feet.	Brazil(i) US Idaho(iv) Puerto Rico(ix)	82 1967 253

(Elderly) 0.5m looks like an elderly man	France(vii) Australia(viii)	138 59
(Reptilian 1) 1.2m oval head, bulging eyes	Puerto Rico (ix)	203
(Reptilian 2) 1.2m large head, bulging eyes, horn on head. Flying	Puerto Rico(ix)	263
(Hairy) Standing about 1.0m - 1.4m high, this person has a round bald head with cat-like eyes and a hairy body which can be brown, gray or black Huge pointy ears. Sightings with 3 fingers	Spain(iii) Switzerland(v) Brazil(i) US Idaho(iv) France(vii) Australia(viii) Puerto Rico(ix)	53 246 24 1968 94 281 204
Angel Reported as 0.5m tall, this appears to be covered in a white dress. It is regularly taken for an angel or the Virgin Mary	Spain(iii) Brazil(i) Holland(v) US Wisconsin(iv) France(vii) Puerto Rico(ix)	173 229 131 1859 19 194

(Uniped) These single-legged creatures work together in pairs to move about. They are about 1.35m high and have been seen in groups of up to 15 individuals, moving like a centipede.	Poland(v)	218
(Ruff) Standing 1.1 – 1.3m, this person has a bald round head and a ruff around its neck	Germany(v)	99
(Electric Man) 75cms. Head and body covered with spiky projections. Blue-green	US Ohio(iv)	1973
Michelin Man Appears bulbous with rolls around its body. It is reported to be between 1.2 and 1.3 m tall. Globe head.	Belgium(v) Holland(v) France(vii)	44 140 260
Monks Clad either in black or white robes, these 0.5m to 1.0m creatures also wear cowls. They always walk one after another in a line. White taller	Germany(v) Holland(v) France(vii)	120 135 294

(Extra Fingers) 90cms. Red/blue skin, white hair, goatee beard long arms, extra fingers	US Indiana(iv) Puerto Rico(ix)	1964 294
(Long Nose and Ears) These have been reported at 1.5m high. They have extremely long noses in proportion to their height.	Holland(v) Estonia(v) Brazil(i) US Kansas(iv) France(vii) France(vii) Australia(viii)	137 58 109 1954 44 178 61
(Helmet) 1m20, moved clumsily. Helmet with facemask and hose to chest.	US Indiana(iv) France(vii)	1973 137
(Astronaut) At about 1.2m, these visitors wear what looks like a diving helmet with a face-plate and tubes connected to it. They seemed to have cat's eyes.	Brazil(i) Belgium(v) US Wyoming(iv) France(vii) Australia(viii)	209 39 1967 255 160
(TV Man) Standing at 1.0m, these Fae wear helmets with antennae attached, sometimes with visors	Brazil(i) France(vii) Puerto Rico(ix) Australia(viii)	166 85 217 169

(Sunburnt)	US Indiana(iv)	1927
60cms, blond hair, pink skin,	US Michigan(iv)	1958
waddles, long arms red hair,	France(vii)	136
pointy ears	Australia(viii)	147
Jenglot	Indonesia (vi)	105
20cms Vicious teeth and claws.	US Illenois(iv)	1981
Cyclops	France(vii)	87
1.0m		
(Tiny)	Puerto Rico(ix)	235
45cms, Large head, webbed hands		
Fairy	Australia(viii)	99
15cms clothed in leaves		

f)   DIFFICULTIES IN IDENTIFICATION

I)   General

I have every sympathy for witnesses asked to describe a sighting of a strange creature which they may only have seen for a short time, possibly at night-time. Their description could easily be flavoured with their pre-conceptions, and they could misjudge heights and colours, as well as miss features which might be critical to their classification. Then there may be problems with the translation of the report sighting.

II)   The Curupira

There is a significant discrepancy between its description in the sighting reports, and that of current folk culture.

In the sighting reports it is described as 60cms tall, with red hair, bulbous eyes, and a skin that looks like a bad case of acne, and claimed to have backward-facing feet.

In popular culture, as described by Google's and Wikipedia's AIs, there is no mention of the bulbous eyes or the acne, but it has kept the backward facing feet. Otherwise he is shown as a handsome youth with flaming red hair.

I have a major problem with the idea of backward-facing human feet. I think this must be a mis-interpretation of some sort of "Other" feet. "Others" have been reported with Human-like feet, hooves, elephantine feet, no feet at all, and simple lumps at the ends of their legs. Equally, there are no Earth creatures which have evolved that way.
Ergonomically, it just seems totally impracticable, and not a survival trait.

Even so, there is a being in Ghana called the Asamanukpai[60], which is also described as having backward-facing feet.

Both AIs flatly deny the presence of acne, suggesting that this association has resulted from the principal acne cure in South America being called Curupira. They do no even consider that they might be confusing cause and effect. The cure could be named after the creature, rather than the other waay round.

This is an example of why these two AIs can annoy me.

III) A Creature with a Trunk

There were a total of 7 reported sightings of Fae with long noses. Some described the nose as being like an anteater's head.

Otherwise it is described as between 1.2m and 1.5m tall, with a hairy body. It has a large head which was described as heart-shaped. The eyes were varionsly described as large, oval and oriental. I suppose these could all be true at once.

This description points to an entity described in Brazilian folklore[61] as a Capelobo.

IV) Small Grays

It is the Small Gray which poses the main difficulty. They are not recorded in these lists because they are present in every single country which I have surveyed, and they occur repeatedly in the Costa Rica (ix) sightings list.

There are at least 4 varieties, causing great confusion. Three are much the same height, with a slender body, thin and long arms and legs, and the distinctive large pear-shaped head with huge black wrap-around eyes.. However, then the problems start:

Type A The classic Small Gray has no ears, tiny mouth and nose, uses its telepathic abilities to abduct Humans, and performs unpleasant medical procedures on them

Type B Then there is the timid version of the same creature which can be scared away if surprised. It squeels, runs away and doesn't use its telepathic abilities, if it has them.

Type C Then there is a similar version but with distinct differences. It has pointy ears and a lipless slit for a mouth. They have also been associated with abductions

Type D Finally there is a much different version with a crest down its back, and wings to fly, and has been described as light green in colour.

In addition to these, there is the Chupacabra, which has been observed[62] being picked up by an UAP, and which uses its fangs and claws far more aggressively. Not for nothing is it known as the Goat Sucker. It could be that it is either another intelligent version of a Small Gray (a **Type E**) or is one of its pets.

To me, it seems probable that Type D and Type E are, in fact, the same creature, particularly as both have been described as having wings. However, whatever the physiology of these creatures may be, they cannot use these wings to fly. As evidenced by the taller creature sometimes called Mothman[63], these wings would have to have been so large, they would have been the primary feature described, and they were not.

I have therefore discounted the Type D entirely and also the possibility that the Chupacabra (Type E) could fly.

It is possible that all the various types of Small Gray need blood as part of their diet, and they exist by drawing blood from cattle, horses and sheep around the world – what we know as an animal mutilation. On this basis, could the tiny Jenglot of Java also be related to the Small Grays? (a **Type F**).

One intreaguing possibility is that the Small Grays are not as strongly telepathic as claimed, and can only achieve mental contact with Humans when actually touching them, Type As are frequently described as carrying an object such as a rod or a sphere. Could it be that these are technological devices which amplify their telepathy and enable them to overpower their victims, delivering the often quoted lie "We won't hurt you".

In this case Type A and Type B are actually identical, except that Type Bs are Type As which have been encountered when not carrying their mental amplifiers.

Another possibility is that Type A and Type B are simply different genders, although it is not clear whether the female is deadlier than the male, or vice versa.

Figure 5 shows four possible types of Gray: Types A and B (Small Gray), Type E (Chupacabra) , Type F (Jenglot) and a Tall Gray.

V)   The Hairy Fae

Looking at the reports in section d of this chapter, the one thing that stands out is there are at least 3 sorts of hairy Fae.
- Beings with hair over their whole body and their head, 1.0m tall with cat-like eyes. Some of these wear helmets with antennae.
- Beings with hair over their whole body, apart from their bald heads. These are 1.2m tall, and have protruding eyes.
- Beings with a thin and hairy body, standing 1.4m tall and with long arms.

There have been sufficient sightings of the first two types of these beings to justify putting them on the list of probable

Fae. We would need more sightings to justify adding the third being.

What we cannot do, at this stage, is determine whether they fit onto my original list of British Isles sightings, or whether they are a totally different species of Fae. Examples of hairy Fae are shown in Appendix 2.

When The Fae Came

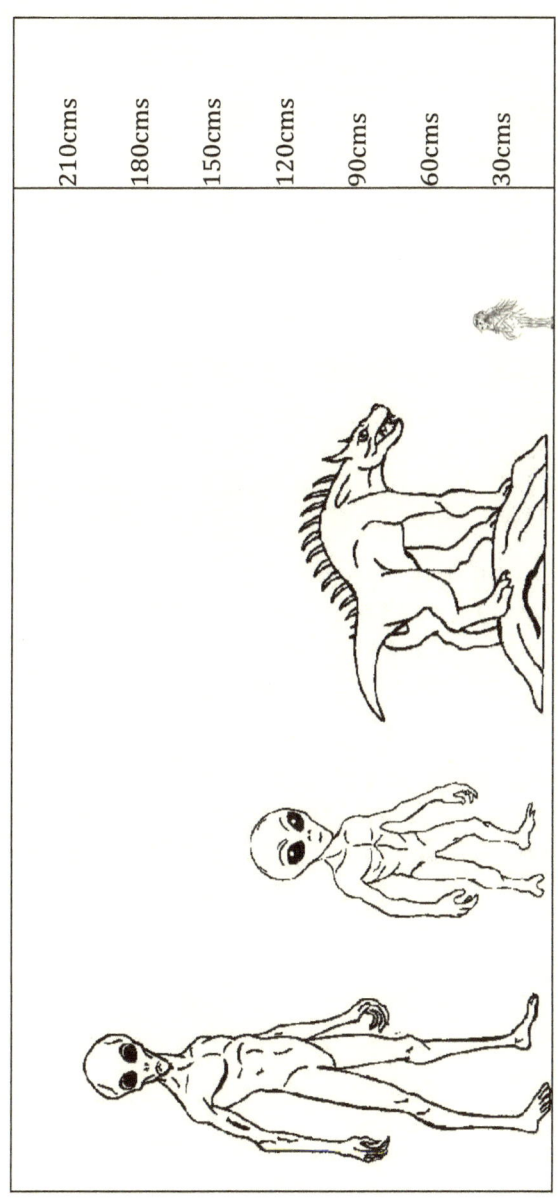

Figure 5 Examples of Gray Aliens
Tall Gray, Small Gray, Chupacabra & Jenglot

VI)   Little Green Men

In the lists in sections c, d and e above, it seems reasonable to exclude green Dwarfs, as the creatures identifisd as such are probably not Dwarfs at all. Too often the word dwarf is used simply to describe any small creature, whilst Dwarf should really only be used for our bearded friends.

In the British Isles, Fae wearing green clothing are often described as "Little Green Men", and this probably occurs in many other places. In Mitrovic's sighting lists, in many of the sightings of "Others" of all types, the being is described as wearing a skin-tight, coloured, one-piece suit.

We need to be certain that the being reported in a sighting actually has green skin, and not green clothing. They must be naked, or the witness needs to have seen their face. We need to go back to the original sighting reports, to see what evidence there is:

The evidence is sufficient to convince me that the two species described as "Frog-like" are clearly not green. The variation in colour described for Type 2 and Type 3 Green Men, certainly casts doubt on their claim to be "Little Green Men".

When The Fae Came

OTHER	WHERE	DESCRIPTION	CONCLUSION
Type 1	Brazil	Green Green Green	Little Green Man
Type 2	Estonia France Spain	Brown-green Black Green Colour not noted Greenish Bright Green	Probably Not Green Man Rename <u>Triangle</u>
Type 3	France Michigan Puerto Rico Spain	Gray Green Luminous Green Emerald Green	Possibly Green Man
Type 4	France Ohio Spain	Green Dark Greenish Face Olive Green	Little Green Man
Type 5	Iowa Indiana	Pale Green Face Bright Green	Little Green Man
Frog 1	Ohio Wyoming	Grayish Gray	Not Green
Frog 2	Illinois Puerto Rico	Brown Skin Colour not noted	Not Green

Bright green or luminous green seem inappropriate colours in Spain. Creatures initially use their skin colour to hide in their environment, even if they later chose to stand out in it by dressing in more flamboyant colours. Bright green creatures would not have easily chosen to live in a Spanish landscape. Their instincts would have fought against that.

It must be remembered that there is a tradition of Reptilian[64] people amongst the recorded sightings so, although they are generally described as tall, there is no reason why some, if not all, of the little green men should not be short Reptiles.

Examples of "Little Green Men" are shown in Appendix 2.

Chapter 5
THEY'RE EVERYWHERE

a) <u>GENERAL</u>

The selection of locations which I used to sample for the presence of Fae was not really made on any scientific basis, other than my interest in assessing how wide-ranging they were.

I elected to start with the British Isles on the basis that it was a well-studied and recorded area, with its maritime constraints limiting casual immigration of other species of Fae.

European locations were an obvious next step, so I selected France and Spain as also being well-studied, and Belgium, Estonia, Germany, Holland, Poland and Switzerland where the records were less intensive.

I needed to know whether the oceans limited their spread in any way so, in the Americas, I selected the more sparsely populated areas in Central United States in the north, Puerto

Rico – a known hotspot - in the Caribbean, and Brazil in the south.

Finally, I selected Australia to see if there were any limits to them.

b) <u>THE RACING CERTAINTIES</u>

Amongst the Fae listed in Chapter 4 sections c, d and e above, there are some whose identity stands out as almost certainly credible:

Curupira 0.6m	This creature's face is easily recognizable, and its dispersed sightings make it highly probable.
(Hairy) 1.0 – 1.4m	These creatures have turned up practically everywhere. There are probably 3 races, although only 2 stand out in this survey.
(Angel) 0.5m	This creature has turned up practically everywhere
Michelin Man 1.2m	This is almost famous, but easy to confuse with the astronaut below.
(Astronaut) 1.2m	There may be more than one species that dresses like this, but how can we tell?
(TV Man) 1.0m	Antenna on the helmet make this one stand out

When The Fae Came

(Octopus) 1.2m	Again this "Other" cannot be mistaken for anything else. Do you include his arms when measuring his height?
(Frog 1) 1.0m	Its thin mouth suggests this might be reptilian in origin, even if not a Frog.
(Frog 2) 0.45m	Being striped green and brown makes this stand out.
Triangle 0.75m	Seen in Europe, where it stands out because of its bright colours. Probably not green skin.

These 10 species are almost certainly genuine Fae, together with those which I singled our previously in Chapter 4:

- Capelobo
- Chupacabra
- Jenglot
- Type 1 Little Green Man
- Type 3 Little Green Man
- Type 4 Little Green Man
- Type 5 Little Green Man

I have deliberately excluded the Small Grays again. Even so, it never ceases to amaze me the variety of physiologies present in this category, particularly when combined with the varieties which I have already listed in Chapters 2 and 3. There are probably many more.

c) **THE RUNNERS-UP**

There are a number of species where there have been a couple of sightings, but not sufficiently consistent to be sure that they are a single species. These include:

- Black Owl
- Short man with ancient face
- Insect Man
- 6 fingered man

These could be re-classified given further sightings.

d) **THE ALSO-RANS**

Apart from those Fae listed above, there are all the other species, listed in sections Chapter 4, sections c, d and e, where there are only single sightings. These are:

- Various reptilians
- Bald with a ruff
- Spiky body
- Cyclops

e) **LIMITATIONS**

In my work, I only surveyed sightings in 10 areas, and it is just a matter of luck whether I picked areas where a particular species is prevalent. I can be certain that I have missed some species of Fae completely.

In Australia, sighting records concentrated on the high number of UAP over-flights, together with Cryptids such as the Yowie (Yeti) and the Thylacines (Tasmanian Devils), although there were a few sightings of Fae such as Fairies, people with antennae on their heads and the ubiquitous Hairy species.

In North America, many of the reported sightings of "Others" were of Bigfoot (Yeti) and Small Grays, but otherwise there were similarities with European sightings.

As expected, Puerto Rico provided a large diversity of Fae sightings to go with a massive influx of Small Grays. The same was happening in Brazil, but at a slightly later date.

Chapter 6
WHY ARE THEY HERE?

a) ELVES & THE LITTLE PEOPLE

Even if that Fae landing in Ireland[65], in 1,700 BCE, was the first ever, it still means that they have been here for over 3,700 years. At the other extreme, we have records of a landing of a tiny species of Fae in Malaysia in the 1960s[66].

There have probably been landings at intervals between these two dates, and before then as well. We can easily go back to the end of the last Ice-Age, about 10,000 BCE. I doubt that the earth would have been as attractive for them before then, given their reported lifestyles.

Even though they have been here a very long time, they have left a remarkably small footprint on the planet. There are a large number of ancient pyramids around the world, but these were probably not constructed by them, or they would likely still be occupying them today.

It would appear that they are trying not to be noticed.

In my previous works[67], I have developed the idea that all these "Others" on Earth are setting up small colonies on other planets to ensure their own species' survival. The universe is a dangerous place, with some predatory species that look on other species as a source of food or slaves. One solution is to set up small colonies elsewhere, which may remain undetected should their home planet be destroyed or conquered.

It is probable that the Fae are doing precisely this. Hence their low profile. Our use of atomic weapons must give them cause for concern, as we are signalling that we are here to all and sundry.

In general they seem friendly to Humans and some types, such as Brownies, actively seek to work with us.

b) <u>GOBLINS</u>

In some ways, these are the most enigmatic. Most types of Goblin seem hostile to Humans, as though they feel they are the true owners of Earth. Maybe it is in their nature to resent competition for the land. I imagine that they are colonists in the same way as the Elves, but it is not clear where they are based.

Perhaps they are instinctively loners, and have simply spread out around the planet.

Anomalies such as Hobgoblins and Duende do exist, and they try to work along with Humans.

c) <u>**DWARFS**</u>

It is thought they came to Earth some 5,000 years ago, when they landed in Anatolia in Turkey, and began teaching Humans about mining and metal-working[68]. They led us through the Bronze Age and the Iron Age, spreading around the world as they did so. It now seems that they are withdrawing to their colonies in Italy and South America, and leaving us to take things forward.

However, it is unlikely they will give up their love of metallurgy and mining, and it is possible that they assist "Others" in excavating underground colonies. They have been blamed for the combination of earthquakes and loud booms recorded around the world.

d) <u>**OTHER FAE**</u>

Most of these have not been observed in sufficient detail to see what they do outside their colonies. Some seem to be studying us, and trying to reduce our usage of atomic power in all its forms.

Appendix 2 shows illustrations of some Fae.

Chapter 7
SUMMARY & CONCLUSIONS

a) <u>SUMMARY</u>

I found it relatively straightforward to convince myself of the existence of some UAPs, and many "Other" sightings around the world. In particular, it is obvious that those that claim UAPs are US experimental aircraft, are deluding themselves about the number of sightings outside the US. Similarly the sightings of "Others" are not just of Bigfoot (Yeti) and Small Grays.

There is confusion about "Others" who are shorter than the Small Grays. Some call the whole alleged family of magical beings elves with fairies as a type of elf, and some call them fairies with elves as a type of fairy. I decided to call them The Fae.

As the Fae display the same capabilities as taller "Others" - invisibility, the ability to fly, telepathy and some mind control, I have concluded that they are not magical, but simply capable of a high level of technology.

When The Fae Came

Analysis of British Isles sightings of Fae showed that there were 4 main species present there: Fairies, Dwarfs, Elves and Goblins, although they do get called different names in different parts of the Isles less than 160km (100 miles) apart.

This problem became particularly apparent when I looked at reference books claiming to list the names of Fae in other parts of the world. There were very few names in common, apart from those in Russia. If I included in my survey all the similar sightings in these books, there would be every chance of double-counting or even triple-counting if adjoining countries called the one creature by different names. The best I could do was to look for correlation with the species identified in the British Isles. I could not look for extra Fae species beyond these four.

I therefore selected a group of 10 regions and used the various reports by George Mitrovic which list all the claimed sightings in those areas, up to the year 2000. There were some species of Fae whose sightings were world-wide and, at the other extreme, some were single sightings. I elected to identify all species shorter than Grays, and with 3 or more reported sightings, as Fae.

There were several species which posed problems. A creature with a long trunk-like nose was sometimes miss-identified as having breathing tubes and a face-mask. There were a number of types of "Little Green Men", which needed separating. Some of these were such bright green that I decided they were probably wearing bright green clothing, and I couldn't know what colour their skin was.

Finally, I tried to determine what these creatures were doing on Earth. I was certain that they were colonists of one sort or another but, beyond that, I could only be certain about some of the Elves and the Dwarfs. Hopefully, more information will come to light as time goes on.

b) <u>CONCLUSIONS</u>

Overall, I have identified a total of 22 species which are highly likely to be Fae, with another 4 which might be identified as such, if further areas were to be surveyed. I have certainly missed some species entirely, just because I didn't pick the areas which they inhabit. I found no mention, for example of the Djinn, although many are convinced that these are "Others" from the Islamic world.

I haven't managed to identify any further Fae colony locations beyond those which I started with, in Mexico, Indonesia, Malaysia, Italy and New Brunswick.

There are many UAP hot spots around the world, but the problem is to identify why these exist.

Some sites are near to important military or research sites. For example, the "Falkirk Triangle" in central Scotland is close to the UK nuclear missile submarine facilities.

Other sites are claimed by taller "Others" to be their facilities. For example the probable site off Argentina, in or near the Mar Del Plata Canyon, is claimed by the so-called Tall Whites, beings which may be from Betelgeuse or Arcturus.[69]

However, there is still the possibility that some of these facilities are shared between "Other" species. This is highly likely in the case of the Canary Islands hot spot, where many varieties of "Other" have been sighted in such a relatively small area. This could also be the case in Hawaii.

It is clear that the Dwarfs have bases in Brazil and Sweden as well as the colony in Italy, and possibly bases in Malaysia and Australia. It is noticeable that fewer Dwarfs can be seen these days as Humans become more competent in mining and metallurgy. The Dwarfs are probably leaving us to it, and returning to their Earth colonies.

The Pixie and the Fairy, which both have wings, may belong to a separate species of Fae entirely. The Hairy Pixie is probably not really a Pixie as it has antennae, so it is likely to belong to yet another species.

Some types of Elves and Goblins – Brownies and Hobgoblins – actively seek Human company. These get given different names from country to country.

In general, though, the Fae continue in their main objective – that of staying out of sight, in order to preserve their species.

APPENDIX 1

Table 1 Number of sightings, in the different areas of the British Isles, of beings of various heights

When The Fae Came

REGION		HEIGHT															cms	SUM
		10	20	30	40	50	60	70	80	90	100	110	120	130	140	150	imperial	
		4"	8"	1'0"	1'4"	1'8"	2'0"	2'4"	2'8"	3'0"	3'3"	3'7"	3'11"	4'3"	4'7"	4'11"		
SOUTH WEST			1	5		3	4	1		4	1						Up to 1945	19
					1			1		3	1		2			3	1946-1975	11
		2							3	1			2		1	7	1976-2000	16
SOUTH EAST				1						3							Up to 1945	4
			1		1	1				1				3			1946-1975	6
					1		3			1						4	1976-2000	10
LONDON					1	2											Up to 1945	3
				1													1946-1975	1
			1							3	1	1	2			4	1976-2000	12
EAST of ENGLAND				3			1	1		3	2	1					Up to 1945	8
			1		1		1	1		1			1			1	1946-1975	5
							1			3	1						1976-2000	4
EAST MIDLANDS				3		1	1	1			2						Up to 1945	8
			1														1946-1975	1
										3				1		1	1976-2000	5
WEST MIDLANDS														1			Up to 1945	1
							2		2		1	3					1946-1975	8
						3					1					3	1976-2000	9
SOUTH WALES										1	1				2		Up to 1945	4
																1	1946-1975	1
																3	1976-2000	3

68

When The Fae Came

REGION	Period	| 10 4"	20 8"	30 10"	40 1'4"	50 1'8"	60 2'0"	70 2'4"	80 2'8"	90 3'0"	100 3'3"	110 3'7"	120 3'11"	130 4'3"	140 4'7"	150 4'11"	SUM
NORTH WALES	Up to 1945	1									4						5
	1946-1975									2	1					1	4
	1976-2000		1								1	2	2		1	2	8
YORKSHIRE	Up to 1945		1	1	1					1							4
	1946-1975			1		2											2
	1976-2000						2			1		2	4			6	16
NORTH EAST	Up to 1945																0
	1946-1975	1	1		1	1											4
	1976-2000						1			2			1			4	10
NORTH WEST	Up to 1945			1			3	1		2							7
	1946-1975		1				2			3	1					3	10
	1976-2000	1			1	1	1	1		1		1	5			2	12
SCOTLAND	Up to 1945				1		6		1	2	2		1			2	12
	1946-1975		1		1					2	1		1	1		1	6
	1976-2000		1				1			2	1		2	1		8	15
IRELAND	Up to 1945		1	11			2	1		2		2				3	23
	1946-1975			2	1		2	1		1							7
	1976-2000		1			1	3			1	1					8	15
ISLE of MAN	Up to 1945			1		2				6							9
	1946-1975																0
	1976-2000									1							1
TOTALS		5	13	31	9	19	39	8	6	48	20	8	26	9	4	66	311

When The Fae Came

When The Fae Came

APPENDIX 2

Examples of various types of Fae

FAE	PAGE
The Alux	72
The Brownie	73
The Capelobo	74
The Chupacabra	75
The Curupira	76
The Dwarf	77
The Elf	78
The Fairy	79
The Goblin	80
The Green Goblin	81
The Hairy Goblin	82
The Slavic Goblin	83
The Hobgoblin	84
The Jenglot	85
The Kappa	86
The Leprechaun	87
The Lutin	88
The Menehunes	89
The Michelin Man / Astronaut	90
The Pixie	91
The Hairy Pixie	92
The Reptilian	93
The Small Gray	94
The Tall Gray	95

THE ALUX

These are called Chaneque by the local Mexicans, and are considered relatively friendly.

They are just 30cms tall, and look like miniature humans. They just fall within the general definition of Little People, but are clearly different from Fairies.

It is claimed that they are the descendents of the occupants of a spacecraft which crashed in Mexico thousands of years ago, and that they live in a colony constructed under Lake Chapala, which is near to Guadalajara.

In terms of surface area, the lake is very large, at 75km x 25km, but shallow with a maximum depth of 9m.

The Alux have now spread throughout Central America, and have been there long enough to have entered into local folklore.

They have a reputation for stealing children, to the great distress of their parents, but are known to look after them, play with them, feed them and return them after five days.

THE BROWNIE

This is a member of the Elf family who has chosen to move into a Human dwelling and to assist with the housekeeping. They are typically 75cms tall, and are generally dressed in rags.

They prefer not to be observed, so they generally work at night performing all manner of chores, including farm labouring.

They do not like thanks, or gifts of clothing, and they may disappear in a huff if that happens. If food is left out for them, it is best to pretend it was accidental.

They are similar in many ways to Hobgoblins, Duende in Spain and Chile, the Slavic Skritek, Heinzelmämmchen in Germany, Wichlin of Austria, Swedish Nissen and the Swiss Servan. Some may even be the same species.

THE CAPELOBO

These Fae are described as 1.2m to 1.5m tall. They are part man with a hairy body, and the head of a giant anteater.

However it doesn't eat ants. It kills small mammals and drinks their blood and eats their brains.

It lives in the rainforests, and it can only be killed with a rifle shot that pierces its belly button.

THE CHUPACABRA

These were first sighted in 1995, in Puerto Rico, and sightings have now spread into Florida and the north of South America.

Inevitably, its existence has been challenged, amidst claims that it looks exactly like a monster from the film Species, and it is just a fantasy. Then it is a very mobile fantasy.

It has been described as doglike with massive rear quarters like a kangaroo, and 1.0m to 1.2m high with spines down its back. Most significantly, it is alleged that it attacks livestock, draining their blood to drink it. Its name is the Portuguese for "Goat Sucker".

There is also a question about its level of intelligence. It is suggested that, instead of being one of the Fae, it is merely a pet. It has been seen climbing into a UAP, so it is connected in some way.

The descriptions in the sighting reports in the US seem to differ from those elsewhere. They may be of a different creature. Time will tell.

THE CURUPIRA

The sighting reports describe these Fae as 60cms tall with red hair, bulbous eyes and a skin that looks like a bad case of acne.

Tradition has it that their feet point backwards, so that their tracks confuse hunters. This is questionable, because nothing else has evolved in this way, so it is not exactly a survival trait in practice. However, the fact that you cannot see a foot at the front does not mean it is at the back. It is more likely that they don't have feet at all in the Human sense. Perhaps, they have no feet at all.

Things have become more complicated now, because the premier acne treatment in Brazil is called Curupira, presumably named after the Fae, but some argue that the monster is named after the treatment!

Interestingly, in modern cartoon culture, these Fae have lost their acne and bulbous eyes, just keeping the red hair and backwards-pointing feet.

THE DWARF

Dwarfs are generally between 0.8m and 1.0m high with a sturdy build and wide shoulders. They are known for their beards, and are famous for mining and metalworking skills. Their smaller cousin the Gnome has similar traits but is more passive.

There is no definitive proof that Dwarfs came from Anatolia, or that they began the Bronze Age.

However, it appears to have started there about 3,000 BCE. Indirect evidence comes from ancient Anatolian houses, particularly those at Çatalhöyük, which were too low for humans, closely packed, and had flat roofs that served as streets. In Eastern Iran, there is another village sized for Dwarfs, called Makhunik.

There is a Dwarf colony somewhere near Genoa. Also there is a Dwarf base in Australia, and probably a couple in South America.

In addition to the probable Dwarf landing in Turkey millennia ago, there was almost certainly another more recent one in central South America, probably now with a colony being developed.

The Dwarfs today seem to have withdrawn from Human contact in most places, probably retiring to their colonies. They are really only spotted now in South America.

THE ELF

The definition of Elves seems to be fairly loose. I am going to describe them as slender and graceful with mainly human features but with pointed ears.

Those in Scotland are apparently as tall as Humans, but the tallest seen in the rest of the British Isles are seldom higher than 1.50m.

The Elves landed their spacecraft in Ireland in 1900 BCE, and burned their ships. They ruled there until 1700 BCE, when they were forced, by the invading Celts, to take to living underground. Since then, they have spread out across the British Isles, and probably much further afield.

Elves are just about ubiquitous. It is interesting how widespread Elves are in North America, where they are probably known to every Native American tribe..

It seems clear that, to be this widespread, Elves have either been on Earth a very long time, or have made multiple landings in addition to the one in Ireland. They seem to be happy to live in the countryside in groups, and sometimes as solitaries as in the case of house Elves. They are reputed to live in caves in the winter, and in temporary sites such as under logs in the summer.

I haven't come across any evidence of their building colonies but perhaps, if they have been here that long, they have finished such things long ago.

THE FAIRY

Many people consider Fairy and Elf as generic terms covering the whole range of Fae. It seems easiest to let the phrase Little People cover Fae under 30cms tall. Of these, only the Fairy has wings as, probably, it is the only one small enough and light enough to be capable of flight on its own.

It is suggested that some Fairies glow in the dark. The one non-magical explanation is that it is based on the same principles that enable UAPs to fly in our atmosphere. A power source is needed, and it has been suggested that power is transmitted through the atmosphere from pyramidal generators such as the one in Mount Denali in Alaska to power these craft. Fairies would only need the capability of drawing on this energy – not an impossible capability. Bio-luminescence is a well-understood phenomenon.

There are nowhere near as many sightings of Little People-sized Fae around the world, as there are of Elves, but they are clearly not all of one species. The Alux and the Jenglot are examples. The Gnome also falls into this category, but is clearly related to the Dwarf.

It could be that the Fairy is related to the Pixie, both having wings, even though the Pixie is too heavy to fly. There is also the French Korrigan, with wasp-like wings.

It is possible that the Fairies do not have the same level of intelligence as the other Fae, and are ruled more by their emotions. They could simply be pets of the Elves.

THE GOBLIN

Goblins are best recognized by their prominent ears which stick out sideways. They are typically 0.6m tall.

They have a reputation for being particularly nasty. Most types of Goblin seem hostile to Humans, as though they feel they are the true owners of Earth. Maybe it is in their nature to resent competition for the land. They are probably colonists in the same way as the Elves are, but it is not clear where they are based.

Perhaps they are instinctively loners, and have simply spread out around the planet.

There seem to be many forms of Goblin, with almost all basically malevolent. However, anomalies such as Hobgoblins and Duende do exist, and these try to work along with Humans.

.

THE GREEN GOBLIN

I'm afraid you'll have to imagine that his face, arms and legs are green. It does have Goblin-like ears.

There seems to be nothing known about their traits or activities.

THE HAIRY GOBLIN

This could be a Duende from Spain or South America, or a Dush from France. These may be different names for the same creature.
It is a house-goblin, but not a welcome one, as it delights in causing trouble.

THE SLAVIC GOBLIN
Known as a Skritek, this is a house-goblin in Slavic areas, much like a Hobgoblin.

In Germany, he is portrayed as a younger person.

It is said to live in the stable or behind the oven. It assists in household chores such as herding animals, sweeping and weaving.

THE HOBGOBLIN

A Hobgoblin is a member of the Goblin group of Fae who is shorter than the traditional Goblin at 0.4m. They prefer not to be exposed to the light of day.

Traditionally is is one of many Fae who have elected to work with Humans and to enter our homes. When they moved in they would live in the hearth and come out at night to perform tasks such as cleaning, making butter or grinding flour. In return they expect respect, and are fed with bread and milk.

At one time, they were the most numerous of the Fae, but modern appliances have driven them out so that they are almost extinct in developed countries.

They can make bad enemies if miss-treated, reverting to more typical Goblin behaviour until they receive an apology or an act of kindness.

They are similar in many ways to Brownies, Duende in Spain and Chile, the Slavic Skritek, Heinzelmämmchen in Germany, Wichlin of Austria, Swedish Nissen and the Swiss Servan. Some may even be the same species.

THE JENGLOT

A Jenglot is a small, mythical creature from Indonesian and Malaysian folklore, resembling a tiny, deformed humanoid doll with long hair and nails. It is typically under 30cms tall, with long claws and sharp teeth.

They are reputed to be blood-suckers.

Bodies of Jenglots are produced from time to time. These are frequently revealed to be elaborate hoaxes using taxidermy or human hair.

However, live specimens have been exhibited and their captor claims to have to feed them with blood on a monthly basis, to keep them healthy

THE KAPPA

The Japanese Kappa is a vampire creature living in ponds. It is 60cms tall, has a tortoise shell-like carapace on its back, and is green with a long nose, round eyes and webbed hands and feet. It smells like fish.

It lives in ponds and lurks there to catch animals when they are drinking. It then drinks their blood.

It is reputed to be surprisingly courteous, honourable and trustworthy, keeping any bargains it may make.

It supposedly only comes out of the water to rape women and to rip out people's livers.

THE LEPRECHAUN

This Irish Elf is the Fae shoemaker. He stands about 75cms high and looks like a little Human apart from his pointy ears.

He is reputed to have a pot of gold, which Humans often try to get from him, but he is very wily. Also, his gold will turn to dead leaves overnight if taken.

He dresses very nattily and is, naturally extremely careful with his shoes, which have very shiny buckles.

It is claimed that they only live in Ireland, so perhaps they consider the Irish areas in the US to be honorary Ireland.

There are, however, similar creatures in other countries: The Kabouter in Holland and Menehuna in Hawaii.

THE LUTIN

These rather capricious Fae, Goblins at heart, are originally from Brittany in France, but they have now moved further afield, being found in parts of the US.

It is claimed that their size can vary widely from a few centimetres up tp about 45cms.

They will guard cattle, in return for milk.. They generally have no permanent home, though some may become house-Lutins. They love to ride horses bare-back, knotting up their manes whilst using their fingers to hang on.

Nain Rouge is the name of a Lutin from Norman mythology, and this name now crops up in Ottawa and Detroit acting as a harbinger of disaster.

By claiming that they can change shape, the idle and lax can blame inanimate objects for their own faults.

When The Fae Came

THE MENEHUNES

These are members of the Dwarf family, living in Hawaii and other Pacific islands. They are very hard workers alleged to have completed large constructions in a single night.

They are typically 1.0m tall with pointed ears, strong, with reddish skin and big eyes and bushy eyebrows.

They are reputed to have already been on the islands when the first Humans arrived in about 2000 BCE. Those in Hawaii were driven back to the mountains of Kauai. It is claimed that they built the Alekoko fishpond, and a number of canals

THE MICHELIN MAN

It is interesting that, amongst the sightings, there are reports of a Michelin Man and an Astronaut. The only difference is probably the witness's familiarity with space suits and tyre advertising

The sightings were widely dispersed: Australia, Belgium, Brazil, France, Holland, France & Wyoming in the US

The descriptions are:
- Appears bulbous with rolls around its body. It is reported to be between 1.2 and 1.3 m tall. Globe head.
- At about 1.2m, these visitors wear what looks like a diving helmet with a face-plate and tubes connected to it. They seemed to have cat's eyes.

Although we can't be certain that all the sightings are of a single race, it is probable. They are short at 1.2m (almost 4'), possibly with vertically slit pupils to their eyes. They haven't been reported to converse with witnesses; they just go around gathering samples of plants and soil.

THE PIXIE

This is a species of Elves from British folklore, about 60cms tall, who often wears green clothing, and looks like a small Human apart for the large ears.

They seem to like leading travellers astray.

They are reported to have small wings, but not that they can fly, and to have heads a bit too large for their bodies. They love dancing

It has been suggested that they are a separate species of Fae, not a Dwarf, Elf, Little People nor Goblin. In that they have wings like a Fairy, even if they can't fly because they are too heavy, perhaps these two are related.

THE HAIRY PIXIE

Known as a Farfadet in France, this has been observed in South-West England

They guard treasure, but also help around the house. Sometimes they are a little mischievous.

This is a rather confusing creature. It probably isn't a Pixie because it doesn't appear to have wings, yet it has antennae, so it cannot be a Dwarf, Elf or Goblin. It must be some other type of Fae.

THE REPTILIAN

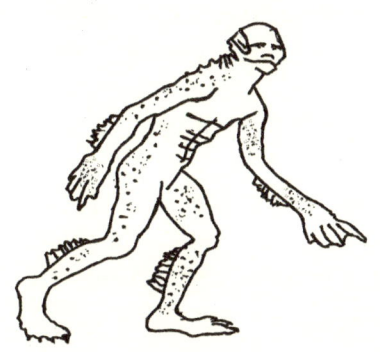

Reptilians claim to be the original inhabitants of Earth, who left before the Humans started spreading round the world, and have now come back.

Again you will need to imagine their whole body green.

They prefer to live underground, and are generally not very friendly to Humans.

They are generally described as tall, but there is no reason why there shouldn't be shorter versions, which would qualify these as Fae. It is possible that these are the "Others", which have been described by some witnesses as Frog-like.

THE SMALL GRAY

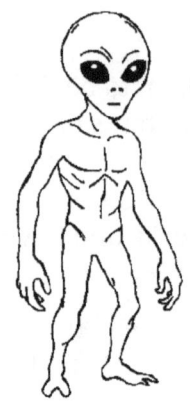

This creature first appeared on Earth in the early 20th century. In 1954, President Eisenhower signed a deal with them to provide futuristic technology in return for a base in the US and permission to take animals and Humans for testing, provided the Humans were returned unharmed. They cynically ignored this agreement once settled in.

They are approximately 1.2m - 1.5m tall, and so are outside my criterion for Fae, but their presence is so significant that they get a mention here. They are gray (surprise!) with heads very much larger than normal, with a pointy chin and very large black eyes which fold around the head. They have tiny ears, nose and mouth. They have long, spindly arms and legs, with three long fingers and a shorter thumb. Hence some witnesses say they have four fingers.

They are telepathic, and use a thought amplifier to enable them to control Humans.

They are notorious for carrying out abductions of both men and women, using the often quoted lie: "I'm not going to hurt you", to pacify their victims. Their examinations are painful, particularly for women who, if they remember, feel violated.

They make women pregnant and then steal their foetus or new-born child. They are also believed to be responsible for the animal mutilations where parts of the animal are taken, together with all their blood. Are they vampires as well?

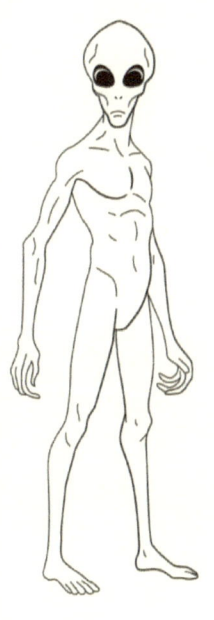

THE TALL GRAY

Standing over 2.0m tall, these only get a mention because they are so often described by witnesses at the same time as the Small Grays.

It is open to dispute whether they or the Small Grays are actually senior, but they do work together.

They look like stretched versions of the Small Grays.

When The Fae Came

Abduction 2, 9, 48, 94
Acne 41, 46, 76
Alux .... viiii, 34, 71, 72, 79
Anatolia 12, 18, 62, 77
Angel 42, 56
Anunnaki 1
Arthur C. Clarke 4
Astronaut 44, 56, 71, 90
Australia ... 21, 28, 32, 37, 40, 42, 44, 45, 56, 58, 66, 77
Austria, ... 16, 37
Backwards-Pointing 76
Bases 2
Belgium 16, 19, 37, 40, 43, 44, 55, 90
Bigfoot 59, 63
Bio-Luminescence 28, 79
Black Forest 20
Blacksmiths 20
Boggart 16, 32
Brazil ... 21, 24, 36, 38, 41, 42, 44, 53, 56, 59, 66
British Isles .. Vi, 4, 9, 10, 11, 12, 14, 17, 21, 22, 27, 32, 37, 55, 64, 67, 78
Bronze Age 13, 18, 62, 77
Brownie v, 10, 16, 17, 22, 32, 61, 66, 71, 73
Canary Islands 66
Çatalhöyük 18, 77
Chaneque. vi, viii, 13, 29, 31,
Chupacabra ... 41, 48, 49, 57, 75
Clay Tablets 1

Cloud Formations 1
Colony .. viii, 2, 13, 20,21, 27, 61, 62,65, 66, 77, 78, 80
Connaught 12
Cowls 8, 43
Cragside 12
Cryptids 7, 58
Cupid 30
Curupira 41, 46, 56, 76
Dactyls 20
Duende 26, 33, 35, 61, 73, 80, 84
Duergar 10, 11, 12, 15
Dwarf v, vi, viii, 3, 10, 11, 12, 15, 16, 17, 18, 19, 20, 21, 39, 52, 54, 64, 65, 66, 71, 77, 79, 89, 91
Earth Ii, 2, 4, 7, 13, 27, 37, 61, 62, 65, 66, 78, 80, 103
Earthquake 2
Elf v, vi, viii, 3, 10, 11, 17, 21, 22, 24, 25, 26, 27, 28, 38, 61, 63, 64, 65, 66, 71, 73, 78, 79, 80, 87, 91
England 6, 10, 11, 16, 103
Estonia . 16, 37, 38, 44, 53, 55
Extra-Terrestrial 1
Fairy ... v. vi, 3, 10, 11, 12, 15, 16, 17, 27, 29, 31, 45, 59, 60, 63, 64, 71,79, 103
Falkirk Triangle 65
Fantasy 1, 5
Folklore 4, 13

France . 19, 24, 29, 33, 37, 38, 39, 41, 42, 43, 44, 45, 53, 55, 90
Genoa..................21, 77
George Mitrovic......6, 16, 36, 37, 64, 103
Germany 16, 19, 24, 37, 43
Gnome ....vi, 3, 10, 17, 18, 19, 20, 21,, 25, 77, 79
Goat Sucker 75
Goblin......v, vi, 10, 12, 16, 20, 39, 32, 33, 34, 35, 38, 61, 64, 66, 71, 80, 84, 88. 91
Greece 20, 25, 33
Harpocrates 30
Hawaii 25, 66, 87, 89
Heinzelmännchen.... 19, 73, 84
Hobgoblin..vi, 10, 11, 12, 17, 32, 35, 61, 66, 71, 73, 84
Holland 16, 25, 37, 42, 43, 44, 55, 90
Humans.vii, 1, 21, 47, 49, 61, 62, 66, 78, 80
Ice-Age 60
Indonesia...............20, 45, 65
Ireland...6, 10, 11, 12, 16, 22, 27, 60, 78, 103
Iris..30
Iron Age 13, 62
Isle Of Man.....................9, 11
Italy.......21, 25, 33, 62, 65, 66
Java ....viii, 48
Jenglot vi, viii, 31, 45, 48, 49, 57, 71, 79, 85

Kappa 33, 71, 86
Kobold...................19, 24
Lake Chapala....................13
Lake Monsters 11
Leprechaun vi, 10, 11, 22, 71, 87
Little Green Man......v, vi, 38, 39, 52, 53, 57, 64
Little People......3, 10, 15, 27, 28, 30, 31
Lutin 33, 71, 88
Magic 3, 4, 7, 8, 12, 16, 63
Makhunik 19, 77
Malaysia.....13, 29, 60, 65, 66
Mar Del Plata Canyon.......65
Marsh Gas 1
Mayan........................viii, 34
Menehune........25, 71, 87, 89
Merpeople 37
Metallurgy 62, 66
Metal-Working 62
Meteor.................................2
Mexico .viii, 13, 25, 29, 37, 65
Michelin Man ..43, 56, 71, 90
Mining ....ii, 11, 62, 66
Monks................................8
Mythical 8
Native American..20, 26, 27, 34, 38, 78
New Brunswick ....viii, 65
Nike...................................30
North America 27, 34, 37, 59
Occam's Razor 3
Officialdom........................1
Pacific Ocean....................26

Pixie . vi, 10, 12, 22, 71, 79, 91
Plesiosaurs 7
Pointed Ears 21, 35, 78
Poland . 16, 21, 25, 29, 34, 37, 40, 43, 55
President Dwight Eisenhower 2
Psyche 30
Puerto Rico 39, 40, 41, 42, 44, 45, 53, 56, 59, 75
Reptilian 42, 54
Scotland 6, 7, 9, 10, 11, 21, 78, 103
Sea Monsters 7
Skeptics 5
Small Gray ... vi, viii, 2, 9, 11, 18, 47, 48, 49, 57, 59, 63, 71, 94
South America .. 7, 21, 62, 77
Spain 16, 26, 35, 36, 38, 39, 40, 42, 53, 54, 55
Submersible Vii

Sumeria 1
Sweden 7, 26, 66
Swiss 19
Switzerland ... 16, 20, 37, 42, 55
Tall Gray 95
Technology 2, 4, 7, 13, 63
Telepaths, ... 4
The Chupacabra 71
The Curupira .... v, 46, 71
Triangle 53, 57
Trows 11
Tuatha Dé Danann 12
Turkey 21, 62, 77
UAP .... v, vii, 1, 2, 4, 5, 9, 12, 13, 21, 28, 41, 48, 58, 63, 65, 75, 79
Wales 6, 10, 11, 103
White Queen 4, 12
Wings, ... 7, 28, 72, 79, 91
Yeti 7, 58, 59, 63

REFERENCES

When The Fae Came

[1] The UFO, ET, Alien Trilogy by Martin Thomas Self-Published 2025
[2] Earth's Alien Syllabus by Martin Thomas. Self-published 2025
[3] Sky People by Ardy Sixkiller Clarke Published by New Page Books 2015
[4] UFOs, Humanoids & Strange Phenomena of Africa, Asia and the Middle East p 199, by George Mitrovic
[5] We Own 29% - ET Has the Rest by Martin Thomas. Self-published 2025
[6] Letters From Mesopotamia A L Oppenheim Pub University of Chicago Press1967
[7] UFO – Friend or Foe Martin Thomas Pub 2025
[8] UFOs: Few answers at rare US Congressional hearing https://www.bbc.co.uk/news/world-us-canada-61474201
[9] Files released on 1974 'Welsh Roswell' https://www.bbc.co.uk/news/uk-wales-10863645
[10] We Own 29% - ET Has the Rest by Martin Thomas. Pub 2025
[11] The Alien Colonisation of Earth's Waterways P184, by Debbie Ziegelmeyer UnXMedia 2021
[12] Alien Archive, The Ultimate Alien Database A-Z p230 by Jacob Sokol 2025
[13] The Fairy Census 1 – Britain & Ireland by S R Young PWCA Books 2023
[14] Field Guide to the Fae Pxxxi, by Nancy Arrowsmith. Pub Llewellyn 2009
[15] Briggs's Dictionary of Fairies. Briggs & Greening Pub Kestrel Books 1997
[16] UFOs, Humanoids and Strange Phenomena of England by George Mitrovic
[17] UFOs, Humanoids and Strange Phenomena of Ireland, Scotland & Wales by George Mitrovic
[18] Alien Archive, The Ultimate Alien Database A-Z p230 by Jacob Sokol 2025
[19] Field Guide to the Fae Pxxxi, by Nancy Arrowsmith. Pub Llewellyn 2009
[20] Briggs's Dictionary of Fairies. Briggs & Greening Pub Kestrel Books 1997
[21] The Little Book of Fairies Mel Barron Pub Octopus 2024
[22] Here Before Us...Giants & Little People Vols 1 & 2 Peter Netzel pub Tired Man Productions 2018
[23] Annals of the Four Masters. www.ireland-information.com
[24] UFOs, Humanoids & Strange Phenomena of Argentine, Chile, Paraguay, Peru and Uruguay p79 by George Mitrovic

When The Fae Came

[25] Earth's Alien Syllabus, by Martin Thomas. 2025
[26] UFOs, Humanoids and Strange Phenomena of Central America, the Caribbean and Mexico as well as the Atlantic Ocean, George Mitrovic.
[27] The Encyclopaedia of Fairies in World Folklore & Mythology by Theresa Bane pub McFarlane & Co 2013
[28] UFOs, Humanoids & Strange Phenomena of Africa, Asia and the Middle East p 199, by George Mitrovic
[29] The Encyclopaedia of Fairies in World Folklore & Mythology by Theresa Bane pub McFarlane & Co 2013
[30] Here Before Us...Giants & Fae Vols 1 & 2 Peter Netzel pub Tired Man Productions 2018
[31] Briggs's Dictionary of Fairies. Briggs & Greening Pub Kestrel Books 1997
[32] The Book of Fairies – Frances Melville Pub Gary Allen Pty 2002
[33] Field Guide to the Fae by Nancy Arrowsmith. Pub Llewellyn Publications 2009
[34] The Little Encyclopaedia of Fairies by Ope Opanike Pub Running Press 2024
[35] Magic, Myth & Mystery of Dwarfs by Virginia Loh-Hagan Cherrylake Publishing 2019
[36] The Fairy Census 1: Britain & Irelad by S R Young PWCA Books & Pamphlets 2023
[37] UFOs, Humanoids and Strange Phenomena of England by George Mitrovic
[38] UFOs, Humanoids and Strange Phenomena of Austria, Belgium, Estonia, Germany, Holland, Kalingrad, Latvia, Lithuania, Luxembourg, Poland & Switzerland by George Mitrovic
[39] UFOs, Humanoids and Strange Phenomena of Andorra, Gibraltar, Spain & Portugal by George Mitrovic
[40] UFOs, Humanoids and Strange Phenomena of Africa, Asia and the Middle East by George Mitrovic
[41] Boggarts, Brownies, Hobs and their Goblin Kin by Stephen Rae Pub The Folklore Press 2025
[42] Fairies by Janet Bord   Dell Publishing 1997
[43] Bronze Age Source of Tin Discovered. The University of Chicago Chronicle 1994
[44] Neolithic Site of Çatalhöyük https://whc.unesco.org/en/list/1405/

[45] Iran's ancient village of Fae. https://www.bbc.co.uk/travel/article/20180109-irans-ancient-village-of-little-people
[46] We ow n 29% - ET Has the Rest by Martin Thomas. Pub 2025
[47] Annals of the Four Masters. www.ireland-information.com
[48] The Dark Pyramid and Violent Nature www.imdb.com/title
[49] Pyramid in Alaska Can Power All of Canada? You-Tube: History 22 Aug 2023
[50] UFOs, Humanoids and Strange phenomena of Bolivia, Brazil, Columbia, Ecuador, Guyana, Suriname and Venezuela by George Mitrovic
[51] Ibid
[52] UFOs, Humanoids and Strange Phenomena of Andorra, Gibraltar, Spain & Portugal by George Mitrovic.
[53] Amazing Encounters with UFOs in Central North America V1& V2 by George Mitrovic
[54] UFOs, Humanoids and Strange Phenomena of Austria, Belgium, Estonia, Germany, Holland, Kalingrad, Latvia, Lithuania, Luxembourg, Poland & Switzerland by George Mitrovic
[55] UFOs, Humanoids and Strange Phenomena of Africa, Asia and the Middle East by George Mitrovic
[56] UFOs, Humanoids and Strange Phenomena of France by George Mitrovic
[57] Amazing Encounters with Monsters and Other Mysteries of Australia, New Zealand, the Pacific and Antarctica by George Mitrovic
[58] UFOs, Humanoids and Strange phenomena of Central America, the Caribbean and Mexico as well as the Atlantic Ocean. By George Mitrovic.
[59] Mermaids on the Menu by Metatron E L.2025
[60] The Encyclopaedia of Fairies in World Folklore & Mythology by Theresa Bane pub McFarlane & Co 2013
[61] Brazilian Mythology P45 by Billy Wellman. 2024
[62] UFOs, Humanoids and Strange phenomena of Central America, the Caribbean and Mexico as well as the Atlantic Ocean. Page 301 By George Mitrovic.
[63] The Extraterrestrial Species Almanac p25, Craig Campobasso, pub Red Wheel. 2021
[64] UFOs, Humanoids and Strange phenomena of Central America, the Caribbean and Mexico as well as the Atlantic Ocean. By George Mitrovic.

[65] Annals of the Four Masters. www.ireland-information.com
[66] UFOs, Humanoids and Strange Phenomena of Africa, Asia and the Middle East by George Mitrovic
[67] The UFO, ET, Alien Trilogy by Martin Thomas Self-Published 2025
[68] Earth's Alien Syllabus, by Martin Thomas. 2025
[69] Ibid

When The Fae Came

www.ingramcontent.com/pod-product-compliance
Lightning Source LLC
Chambersburg PA
CBHW060459080526